I0100381

Blueprints to Building Your Own Voice-Over Studio.

For less than $500.

By Ian King

© 2020 dotandtittle publishing
eBook ISBN: 978-0-463-34915-1

Hardcover Edition – Copyright 2019 Ian King.
Softcover ISBN: 978-0-473-55950-2
Hardcover ISBN: 978-0-473-55982-3

All rights reserved. No part of this book may be reproduced or transmitted in any form or by any means, electronic or mechanical, including photocopying, recording, or by any information storage and retrieval system, without permission in writing from the copyright owner.

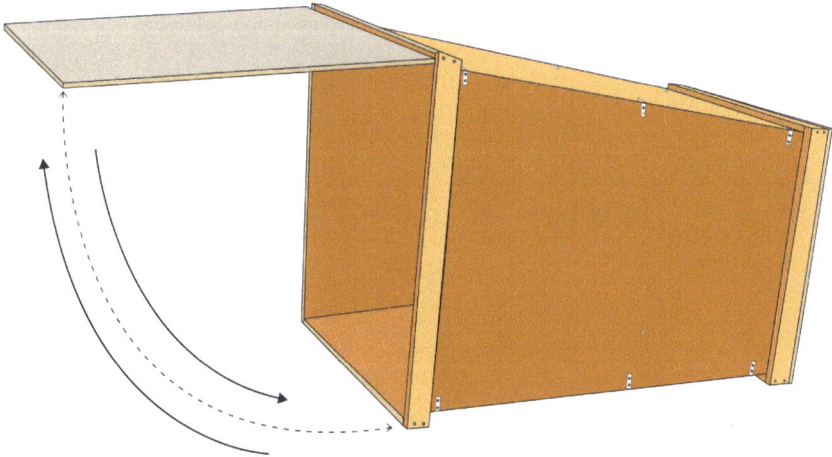

Table of Contents:

WELCOME:

This book is a manual on how to build your own studio for Voice–Over work and narration. It is a technical book, therefore, I suggest that you read it completely through to begin with before you go ahead and build your own studio.

It may sound confusing at first, but stick with it and all the pennies will fall into place for you.

LHS + RHS + Top:

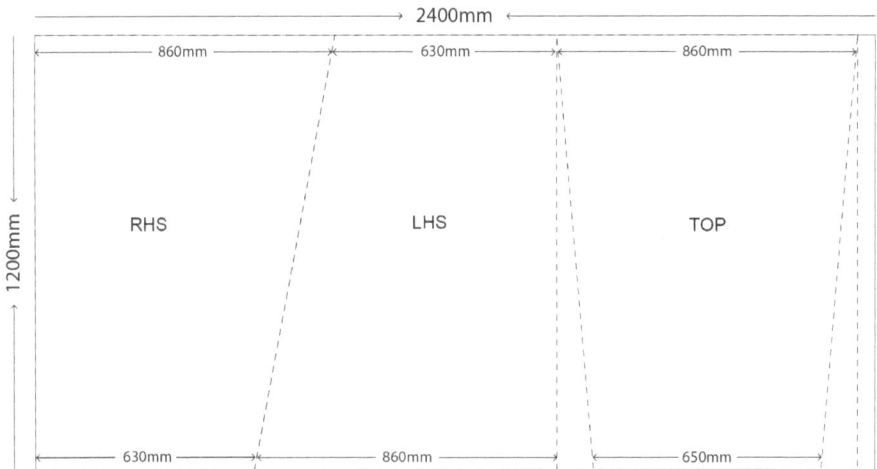

Bottom + Lid + Tapered End Panel:

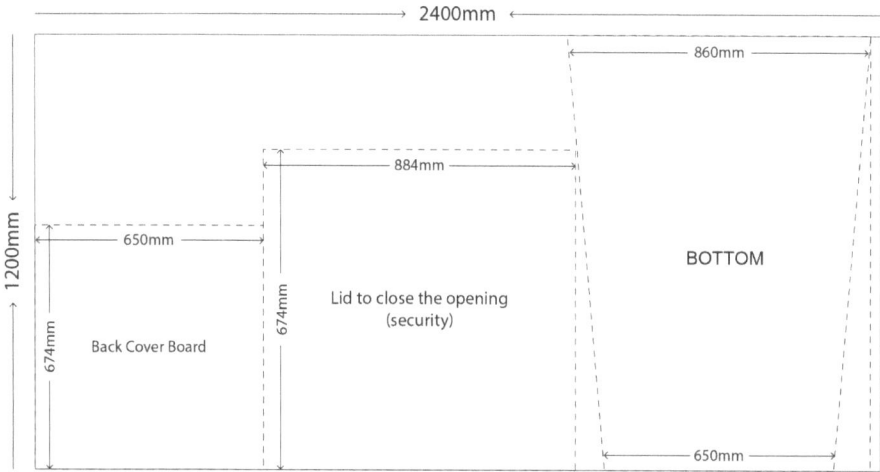

2400mm

1200mm

860mm

884mm

650mm

674mm

674mm

BOTTOM

Lid to close the opening
(security)

Back Cover Board

650mm

Exploded View:

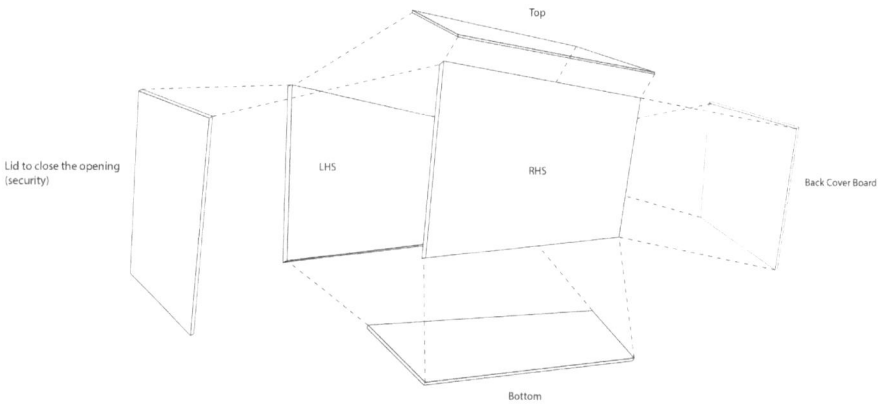

Top

Lid to close the opening
(security)

LHS

RHS

Back Cover Board

Bottom

INTRODUCTION:

There is a world of opportunity a stone's throw away from you; in fact, once you are sitting in front of your PC or LAPTOP, you are staring into a gold field and all you need to do is pick a spot and start digging. You will need tools of course, and the mine where you chose to start harvesting will determine the tools that you will need.

This book is a guide to developing the tools you need for PROFESSIONAL VOICE–OVER WORK. This field is huge! You supply the talent and I'll show you what to do with it. But there is no point in digging these fields if your talent is not here. You'll obviously need good articulation and confidence in your vocal abilities. You'll probably need some proper training too, but this is not what I'm offering. I'm giving you practical guidance on how to build a space to work from, at a very low cost.

The key thing to always bear in mind with voice acting and audio works is to fight the chief enemy of recordings… and that is 'noise'. Electronic humming, fan noise, bumps, bangs, bad weather, echoes and lawn mowers… you name it. So, I've developed a small recording space that gives you plenty of room to gesture wildly within as you articulate that most dramatic audio book, yet you can still get up close and personal with the microphone to create your own take on 'the Voice of God' for that movie trailer you might be cast for. And don't forget you'll need to share that space with some equipment. This Home Studio Space, or Vocal–Booth, is solid and well designed with these considerations in mind.

I'll clarify other things you will need to consider as you read this book and undertake this project, so you know what to expect. My book is filled with photos, plans with specific measurements; detailed instructions and guidance.

First though, is the 'why' we need to consider the enemy of quality recordings and why we must battle it so vehemently. I'm talking about 'that NOISE'!'

NOISE: This comes in many wave–shapes, forms and sizes but, whichever way they come, all of them will destroy your recordings and render them unacceptable.

There are external noises like cars running up and down your street, dishwashers cycling through a load, people talking and babies crying. All these need to be reduced somehow. But try telling your whole street to be silent while you are performing your recordings... Good luck!

Then there are internal noises you need to get rid of too by keeping them out of your recording space: like computer fan noise, humming caused by electricity problems, echo sounds, or boxy–sounding results and so on. Internal noise is a significant concern in recording which is why the Personal Studio is specifically designed to assist you in achieving professionally acceptable results.

This Personal Studio has a minimum of three walls of defence:

1) Insulation – to REPEL external noise

2) Absorption – to, well *absorb* and *treat* internal noise

3) Design – to also reduce internal/external noise, leaving you with a cleaner and clearer result.

I'll expand on these three:

> **i) Insulation:** The 12 mm plywood walls used to construct the studio are lined with an insulating, muffling shield that you can likely get for free, much of the time. Visit some whiteware or brown ware stores (fridges, TV's) and even furniture stores to source sheets of Polystyrene 15–25 mm thick. The bigger and thicker the sheets the better.
> **ii) Absorption:** You cannot cheat on these important elements; for proper absorption, you need proper industry standard Acoustic Foam Panels which are not overly expensive. Egg cartons or mattress foams do not have the right properties to deal effectively with 'noise'

and use of them will lead to a 'boxy' or 'tacky' sound, or a sound that is lifeless and limp, leaving your recordings undesirable to the professional end consumer, and they are your audience who ultimately pay you for your work.

iii) Design: The design of this studio is tapered and of a specific size for various reasons I will briefly touch on. Imagine a pond that you throw a stone into. What happens to the water's surface once the stone has made contact? Of course, it produces little wakes and ripples. These wakes and ripples are very much like sound waves. Notice what happens to these lil' critters as they spread out and eventually hit the edges of the pond. They bounce back, don't they? And then begin travelling in another direction (often, right back into your recording microphone).

When we bring this information into the perspective of what we are talking about here... audio waves! These are the very things we need to eliminate as much as possible.

A tapered shell design has several advantages:

a) It takes up less space and it looks prettier than a boring, square box.

b) The tapered walls are not parallel, thus forcing those lil' bounce–back critters to travel longer and further distances before hitting another surface and therefore changing direction again... this resulting pattern of wave–travel eliminates much of those amplification troughs and peaks of sound–wave intensity.

c) The front (where you will be moving your arms and hands, and operating your equipment) becomes a useful working space, whereas the rear (end) of your studio is neither wasting space nor adding weight to the final product.

Unfortunately, some slapped together, home-recording spaces for recording audio books and voice-over work in general can leave a boxy, irritating noise that can be difficult to listen to over extended periods of time. That is the reason this design is so brilliant! It's not very expensive to build, it does a magnificent job of supporting your voice talent, and it gives you a place to produce an excellent final result.

Next, is a list of things you will need with approximate costs. Everything I have specified below has frugal, and design in mind.

Part One – CORE CONSTRUCTION/PRODUCTS:

Plywood panels: 3 panels: 1200 mm x 2400 mm x 12 mm thickness. Approximately $75

($25 per sheet). You may also use Custom Wood panels instead. They are heavier, but more dense.

Smooth–sawn timber planks: (about two: 70 mm x 25 mm x 1800 mm).

Approximately $20

Acoustic foam panels: 15 panels: 300 mm x 300 mm; plus 4 panels: 500 mm x 500 mm.

You can get these 50 mm, or 30 mm thick; however, I strongly recommend 50 mm.

Approximately $80–$130.

Polystyrene sheets: hopefully for free if you ask around at places such as: whiteware, brown ware and furniture stores to name a few.

Fastenings: screws, hinges, chain, various hooks and brackets: screw–sizes between 15 mm – 40 mm; length of chain (around 8 gauge – some of these items are optional and are only needed if you wish to suspend your studio from the rafters/ceiling), at least 2 metres (2000 mm), depending on the height of your ceiling rafters; flat hinges (2–3) about 80 mm in length; 'S' hooks (8) 10 gauge and 50 mm long;

angle brackets (12) around 50 mm; hasp & staple sets (2) 150 mm; strong pulley wheels (2); eyelet hooks (10) 8 gauge 50 – 60 mm

Around $80

Basic tools: hammer, drill, drill–bits, circular saw, circular drill bit, tape measure, pencil/felt–pen, box–cutter knife, ruler.
Canned spray–on glue: two at approximately $40 for both.

Travel blankets: or thick carpet cut–offs. find these at thrifty/second time around shops – usually very cheap, say $10 each, the thicker the better. Get at least 5.

METHOD:

To start with, make sure you have a descent–sized, covered space (like your garage) to work in. You will also need some timber to run along the floor (100 mm x 50 mm x 2 m), as an elevator to lift the plywood sheets off the floor while cutting them to size.

MEASURE:

See the images of the plywood sheets with the measurements drawn up, ready to be cut out.

Take the first sheet and mark up the two studio side panels (left and right) and the uppermost (top) studio panel (see following diagrams).

Left and Right–side panels:

Measure and mark with a felt pen along the 2400 mm edge. These are your two beginning points: the first point at 860 mm, then at 1510 mm.

Run a line down at 90 degrees. This is the bottom edge of the second–side panel.

Now begin measuring the opposite side (from the same end as your first marks), beginning with 650 mm, then at 1510 mm. The panels are ready to be cut out with the circular-saw.

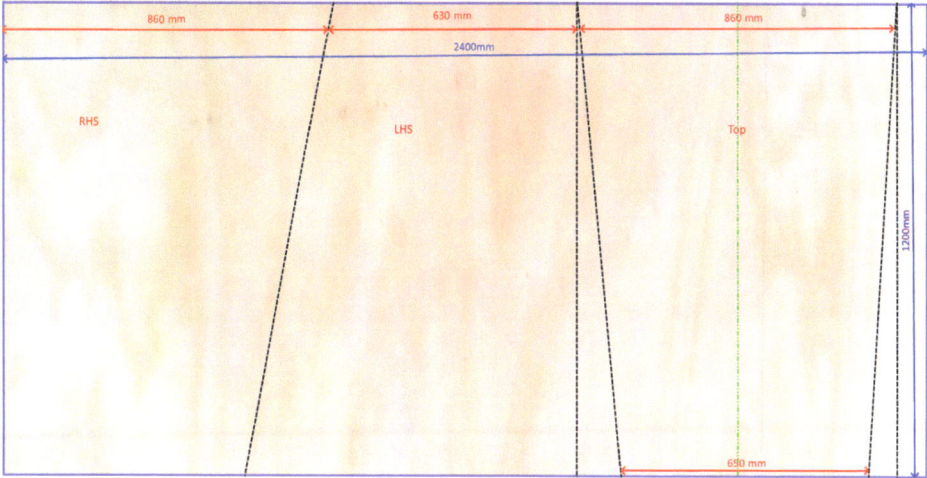

Top Panel:

Continue along the original 2400 mm edge you began marking up. Your next measure mark will be the opening end of the top panel of the studio. Mark up at 860 mm.

Next: find your centre mark, being 430 mm (1/2 of 860 mm) in from each initial marking, left and right.

Once you have found the centre of the opening end of the panel, move to the opposite edge of the plywood to mark out the closed off, smaller tapered end of your studio.

Measuring out from this centre marking: 325 mm left, and then 325 mm right.

Next: mark up the taper of each edge by running a line (left and right) from the open-ended marks, down to the closed-ended marks.

Now the side panels and the top panel are ready to be cut out.

Next: is the second sheet of plywood and we will use this for the bottom panel and the two ends. Marked them up to be cut out.

As with the top panel from the instructions above, mark up the first measurement along a 2400 mm edge of the second sheet of plywood.

The first mark will be at 860 mm, and then find your centre mark again (as per the top panel) and mark this down to the other 2400 mm edge.

Find the other marks of the taper at this opposite side (2400 mm) edge.

Mark out 325 mm left and 325 mm right. This is your underside panel (the bottom).

Next: The small end of the taper will be fixed securely to the body of the studio while the main opening, where your audio narration will actually take place, will in fact be a lid. With this lid you have the ability to:

1) Close up your studio when not in use, for either transporting, or keeping unwelcome guests out (i.e., your kids).

2) Create another noise control barrier. This lid will be elevated above your head while you are using your studio and you will hang some of those blankets in your supply list over it, they will also help to block out external noises and prevent further internal echoes, which actually originate from within the studio as you voice your amazing creations.

Closed, tapered end measurements:

These sizes are slightly large than the body/side panels. This is because the side panels are to be fastened to the inside edge of the top and bottom panels, and therefore the end panels will be another 24 mm wider [12 mm thick plywood panels per side... 12 + 12 = 24 mm].

MEASURE:

Take your last marker, and from your last 860 mm mark and down towards the opposite end, begin marking out the lid at 316 mm [1200 mm - 884 mm = 316 mm; we'll get to that measurement soon].

Note: Place this first marker–point next to the taper mark of the bottom panel. This is to utilise the entire panel of plywood economically; otherwise, you may not have enough room on this sheet to get all your panels cut out. We're trying to use only one sheet for these next panels.

Once you have marked out this point, run your mark down to the bottom of the 2400 mm panel and then measure along the marked out, tapered edge, finding your mark at 884 mm.

Next, mark up [at 90 degrees] the adjacent side to 884 mm and then back to your original marker–point, forming a square. Well done! This is your lid marked out completely, ready for cutting out.

Next, your tapered-end panel:

Go back to the 2400 mm edge (from where you marked out the lid) and measure up (along the marked-out lid panel) to 674 mm [12 mm thick plywood panels... 12 + 12 = 24 mm; thus 630 mm + 24 mm = 674 mm].

Now, run your marker across the 2400 mm edge to 650 mm, then go up the sheet at 90 degrees to 674 mm, and then back to the mark along the lid edge.

Now the tapered, closed off panel is also ready to cut out.

CUTTING OUT:

First, be sure you measure twice and cut once!

Then lay some 100 mm x 50 mm timbers on the ground to create a cutting–out platform which will elevate the plywood off the ground and give the circular saws blade clearance.

Now take your circular saw and start wherever you want, taking care to follow your markings carefully. Be aware of where the circular saw blade is in relation to where its foot marking guides are, for lining up. You don't want to follow the circular saw foot marking guide, only to find you should have used the other foot marking, because now your cut is out by 10 mm!

FIRST TEST ASSEMBLY:

The assembly of the shell can be tricky as some measurements will not be 100% accurate. The measurements will be out of whack by a mm here and a few mm's there, so common sense prevails at this stage.

Gather your screws, angle brackets and the drill. We are going to trial your cutting skills.

First, put the bottom panel on the ground and then the first side panel will be attaching to its cut edge. Lean the side panel up against a wall, then take 3 angle brackets, positioned with one at each end *[not less than 100 mm, from each panels end]*, and one near the centre. Sit the brackets along the bottom panel, making a gap [12 mm] at each. You will slip the side panel into here.

Next, do the same for the other side panel. You will probably need to have something solid for the side panel to lean up against. Fasten the brackets to each of the side panels with screws.

Now, lift the three connected panels onto the larger opening end and stand it up that way. Fasten the brackets onto the bottom panel now with screws and then you can connect the top panel in the same way. This is your basic studio shell, temporarily assembled.

To strengthen this shell as it is, attach the smaller tapered end panel to the end of the studio, closing it off and giving the studio a little rigidity. If you do not put this end panel on the shell will flex and flay about, putting unnecessary strain on the joints and damaging the structure. *(See following diagram).*

PREPARE TIMBER FOR STRENGTH:

Once you have temporarily assembled the studio you can see it taking shape. Now, to prevent this studio from collapsing under movement we need to strengthen both ends. Bands of timber will be used to hold the studio together firmly, preventing it from twisting while moving around. These timber bands will also be used to fix the insulation and soundproofing to your studio.

MEASURE:

At the opening of your studio at its widest point, make a note of its measurement (should be 884 mm along the top timbers and 850 mm along the side timbers). This is the length of your timbers, PLUS add the thickness of the timbers (e.g., 24 mm + 24 mm = **48 mm).

Add this to the lengths of the top and bottom timbers as these will act as the anchor and fastening edges for the side timbers, which will remain at 850 mm.

NOTE: The strengthening timber ends will be cut on angles that run parallel to the shell of the studio, so make sure you measure the top and bottom timbers to the maximum, external sizes of the opening. The end (tapered end panel) will be measured differently as the taper of the studio shell will get longer rather than shorter as you measure up the shell.

After you have measured the 4 x timber edges for the opening end, measure up the timbers for strengthening the tapered end. The difference is that the longest measurement of the four timber lengths will be taken from up and along the shell of the studio and is determined by the width of the timbers that you have purchased.

For example, I have suggested your timbers should be 75 mm wide, but you might have got an excellent deal on 100 mm wide timbers and that's great! The beauty of all of these studio measurements is that all of my suggested measurements can be modified, either by making them all smaller, or by making them all larger; thus, adjusting the size of your final product to either being a smaller studio, or a larger one. Totally custom built!

Tip: All of these measurements can be modified. The trick though, is that you must calculate all variant measurements correctly and consistently; otherwise, you will end up with mistakes and will need to recut, or maybe even start all over again. All my measurements have been carefully calculated and proven to work well.

So, take the width of your timber (whether 75 mm or 100 mm) and measure from the very end of your tapered studio by that measurement, up from the bottom end of the shell of your studio and mark where that comes to.

This will be your longest point of the timber (top and bottom) PLUS the thickness of the timber x 2 (as per opening end's measurements, indicated with '**' previously).

Measuring the two side timbers is easier as all you need to do is take your timbers, and from the bottom (square) edge of the studio side panels, lay the timbers on the side panels and simply mark flush, at the same edge as the side panels top edge. *(See diagram)*

Mark the angle of the studio shell out on the timbers, then cut them.

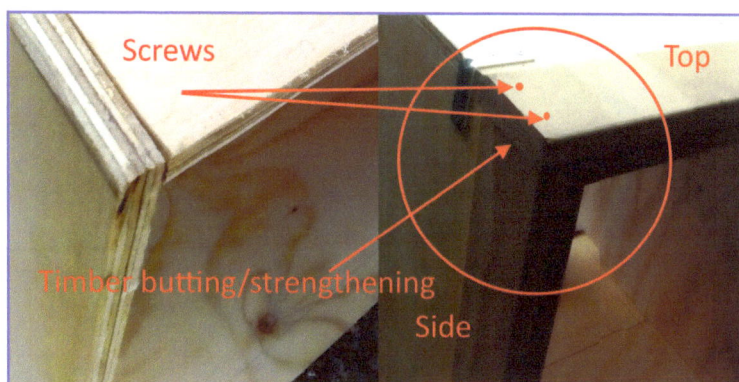

Once the four timber band supports have been marked out and cut (at the appropriate angles), mark where the two screws (per end, per timber) will go and pre–drill these holes so that the timber won't split when joining it all together.

HINGES FOR DOOR:

Now it's time to mark out where the hinges will be fixed to the top timber band at the opening. These hinges will be fixed to the underside of this band, out of sight; that is, the hinges will be screwed to this top band rather than to the shell of the studio.

Find the centre of the top timber band and mark left, then right, to a measurement of around 200 mm each way. This is not critical, provided the weight that they will bare is spread reasonably evenly between the two hinges.

Be aware which is the front side of the band and which is the rear side of the band. Their lengths will be different because of the angles they have been cut at. Fix the hinges to the front, bottom side of the top timber band (the longest edge).

BOTTOM PANEL SPECIAL HOLES:

We have nearly completed this stage of production. The next stage will be insulating your studio and joining it all together. We now need to drill 4 x large holes in the bottom panel of the studio shell to give us a place to run power/USB cables through from various points.

From the outside edges of the panel, mark (from the left edge and the and right edge) to 100 mm, both at the open end and the tapered end, then draw a line from top to bottom along each edge to mark out a centre line to find your place to drill these holes.

Next, measure and mark with a felt pen an '+' from either end along these lines, at two points...

1) 400 mm,

2) 800 mm. Then take your drill with a 50 mm diameter cutter on it and drill out two holes per side,

along each line at the 2 x marks. These holes will accommodate cables and leads you will need to pass through the shell to the outside of the studio. Having four different positions will give you flexibility when passing cables through within your personal set up.

INSULATION – Part Two:

Now, the final pieces of the puzzle. The inside of the shell is well insulated to stop much of the external noises from getting into your recordings and to control, or treat the noise produced from the inside, such as reverberation.

Two more layers and a final layer on top will hold it all in place and make it very presentable.

The first layer, directly under the plywood panels, is polystyrene which you can usually source for free. I sourced mine from a local furniture store. You can also find polystyrene from retailers of brown–ware and white ware (TV's and Fridges/Washers).

Polystyrene sheets are ideally about 15–25 mm thick and should be the largest sheet sizes you can find (here's hoping for sheets around 1200 mm x 900 mm).

Once you have the ideal selection of sheets, mark them up to fit inside your studio shell, taking note that the uppermost polystyrene panels should be worked on first and measured out to fill the entire available surface of the top, or ceiling of your studio and that the top edge of the side panels (left and right), should be supporting these 'ceiling' panels.

The positioning of these sturdy side panels will add to the strength by supporting the inside treatment and will have the strength to uphold the weight of the Acoustic Foams, which will be glued and fastened to them.

POLYSTYRENE PANELS:

There are to be three main sheets in total, and one smaller sheet for the tapered end closure. You will also need more of these sheets to make a couple of backboards which are an added barrier to prevent external noises.

Follow on with (after the ceiling panel) the side panels and mark them up as per the tapered angle of the studio shell. The most logical way to do this is to pull your temporary studio assembly apart and mark the sheets along the edges of the studio wall panels with a felt pen. However, you can also simply remove the end panel of the taper so you can take measurements within the partly assembled studio, via the top end and bottom end. I recommend this way.

Take the height of the opening end (860 mm), also the height of the tapered end (650 mm), and the full length of the studio wall (1200 mm). Lay out your Polystyrene sheets on a flat surface and mark your sheets up according to these measurements.

NOTE: one edge of these panels will be at a 90–degree angle (the portion that runs adjacent to the floor of your studio) while the other edge will be at an angle. This angle will be dependent on your initial studio measurements.

Also, consider the thickness of the polystyrene sheet you have reserved for the top and subtract this measurement from the total height of each side panel [say it's 24 mm thick, then the formula would be... 860 - 24 = 836 mm. 650 - 24 = 626 mm].

Both left and right sheets should be identical so you should only need to measure up one sheet and then copy those markings onto the other sheet. The top panel will be measured the same way; however, this will be marked up to the top opening edge and the bottom opening edge and there will not be a 90–degree angle.

The top panel needs to be tight up against the sides of the studio shell. So, take the top opening measurement (860 mm), the tapered end measurement (650 mm) and the length of this panel (1200 mm).

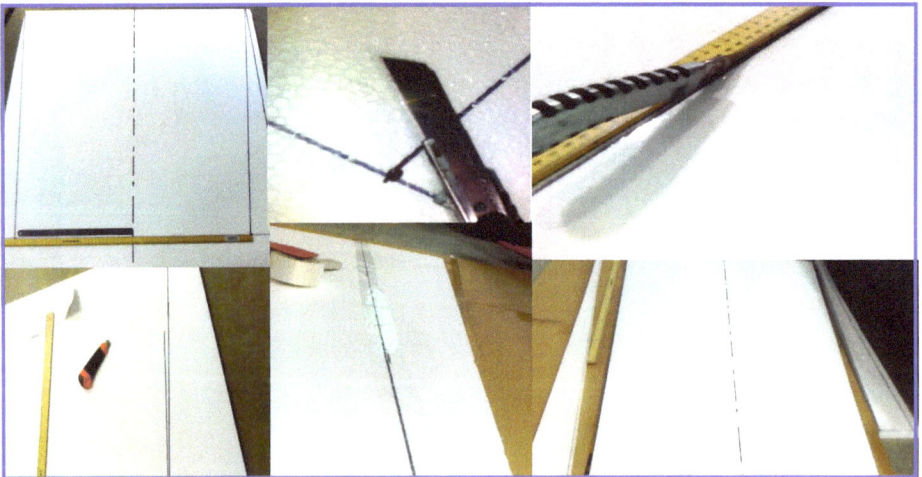

Come back up to the top (opening) width and find the centre... 430 mm from the left and 430 mm from the right (open end), then 325 mm from the left and 325 mm from the right at the tapered end.

Mark this centre line up.

Next is the tapered end, the 'blocked–off', end sheet. Mark this up to the same height as the end of the side panels and the same width as the width of the studio shell's tapered end, less the thickness (x 2) of the side panels.

For example, the thickness of the side panels = 24 mm, so the width of the studio shell will be 650 mm - 48 mm, thus 602 mm x 626 mm.

Once these sheets are marked up, take a sharp box–cutter knife and a long ruler and cut them out.

Tip: **It will make the cutting out easier if you get clear packaging tape and run this along each marked out line for cutting. It will also help if you extend the box–cutter blade to its maximum length and cut with the knife on a low and long angle. Also, put something such as corrugated cardboard or old linoleum underneath the sheets where you will cut since the knife will also cut into anything that the polystyrene is lying on.**

ACOUSTIC FOAM:

It is not a good idea to shortcut this product. Acoustic foam is the industry standard for any professional audio setups and for good reason. Acoustic foams come in various shapes, sizes and thicknesses all for the purpose of absorbing and hinder audio sound waves, to catch and retard these rogue sound waves that threaten to disrupt your perfect audio recording.

I recommend a minimum thickness 50 mm as the thicker the foams the better the results in absorption will be.

Tip: **If you have mail ordered your acoustic foams and find they are a little squashed up or deformed (sometimes they can be crushed for extended periods**

of time and this can cause this effect. It's not usually a fault with the foams), then all you need to do is push the effected panels into a bucket of water and squeeze them a few times, taking care not to tear the foams; once they are thoroughly soaked then pull them out of the bucket and gently squeeze the excess water out, then put them on an airing rack for them to dry over a couple of days.

METHOD:

Select your foams and lay these out on the pre–cut polystyrene sheets to give you an estimated *best economical* layout. It is a good idea to mix and match the direction of how the sound waves will be scattered by laying out the foams ridges adjacent to each other.

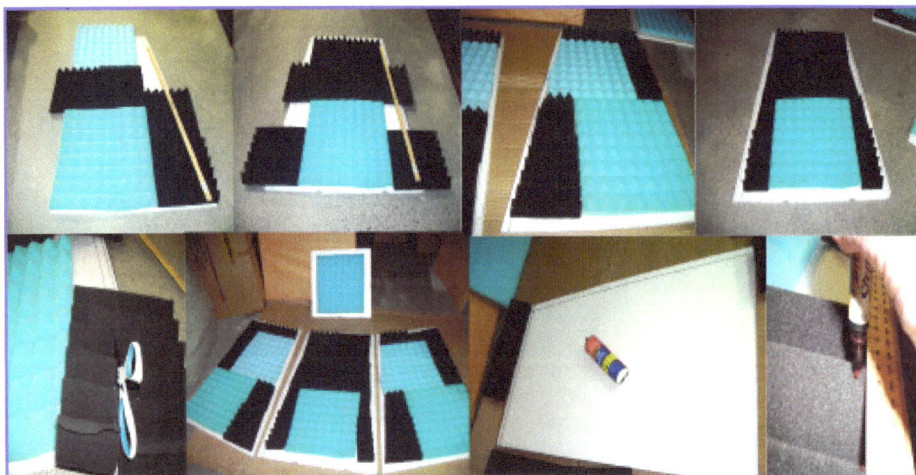

Once you have an idea of where you want to layout the foams then remove them and mark up the polystyrene sheets with guidelines of placement.

Mark–in from each of the edges of the polystyrene, a gap of between 30–50 mm except where the front of the studio will be. This is where you want the foams to be flush with the front opening edges for maximum absorption.

Once the four polystyrene panels have been marked up, it's time to mark out and cut your foams, ready for gluing to the polystyrene panels within your marking guidelines.

With a felt pen and ruler, mark out where you will cut the foams. Use either scissors or the box–cutter. Once you have them cut and assembled on the polystyrene, glue them to the polystyrene.

NOTE: The front section of your panels should not have the foams glued at the very front 150 – 200 mm as the next phase of assemble requires the blankets to be glued under those foams. See the following explanations.

It's a good idea to weigh down the freshly glued foams by pressing on them, using some of your spare plywood and some bricks, or other heavy objects. Leaving them weighted down for at least an hour. This ensures a good bonding of the foams to the polystyrene panels.

BLANKETS:

We are getting close to the final assembly. This phase requires those blankets you have obtained in order to give your studio a nice, tidy and absorbent finish.

The blankets need to be cut to the various sizes for correct placement. Take the thickest blanket and use this to line the floor (bottom panel) of your studio. Lay the blanket under the bottom panel and use this as your stencil.

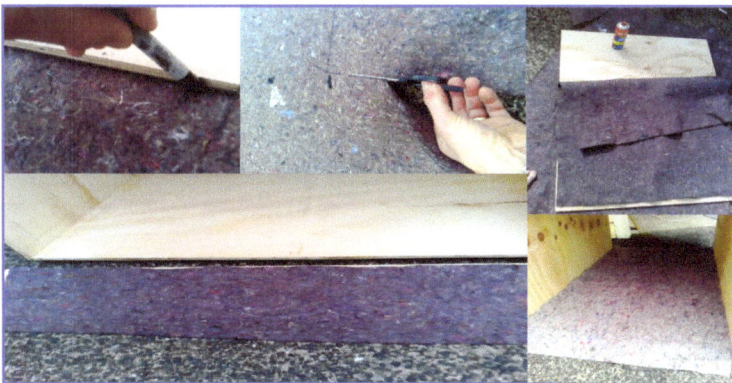

Bear in mind you will need to leave a section/flap to wrap around the front edge (about 150 mm, to be screwed into place with the strengthening timber bands).

The ideal way to use the blankets is to take the width and height of the top and side panels and cut out x 3 rectangular blanket sections by taking the width of the walls, and the ceiling panel, which you will be lining and then extending the adjacent section by 200, to 300 mm. Also, add this 200 mm to 300 mm to the section of blanket you will use for the floor of your studio (as mentioned before, extending it from the widest, or open end of the studio). We will call these four rectangular sections of the blankets, 'flaps'.

Mark out these blanket flap's with a felt pen and then cut them out with scissors. Note that the edges will not be cut square as the side and top panel are all tapered. You can trim these up to size once you are finalising the construction and sealing it all together.

Once these pieces are all cut to size it is time to fully assemble your beautiful creation!

NOTE: Before you assemble this and start the process of adding in the acoustic treated panels, double check that you have drilled the four holes (50 mm diameter) on the underside (bottom) panel. (See earlier diagram.)

ASSEMBLY:

Finally, all your hard work is paying off. Start with your shell, using the angle brackets to fasten all the sides, top and bottom panels together. Making sure you fasten them all from one end to another, to squeeze out any possible 'bowing' of the panels.

Double check that the ends of each panel (particularly the end which will be sealed) are flush with each other. You might need to trim one or two of the open ends (at the opposite end) of the panels because of the taper, which can sometimes throw out your careful measuring. You may also want to trim the closed end top panel, to make it flush with the side panels as the taper will cause the right angle of the 12 mm edge of the plywood to jut out.

Now, take the blanket flaps and position them at their side panels on the outside of the studio, readying them to be

stretched and glued to the inside of the studio. Fasten each with the timber bands surrounding the front opening. But also remember you will be removing the top timber band again when you fix the 'lid' and hinges to the shell. It would make the rest of the job difficult and awkward if you were to fix the lid securely at this stage.

Once you have positioned the flaps to the four walls of the opening and cut them to suite the angles of each wall, then fix the timber bands to the outside edges, pinning the blanket flaps to the walls, one wall at a time. Start along the top edge, screwing the blanket flaps down and then slipping in the side timber bands along each of the two walls, left and right, also pining their blanket flaps to the exterior.

First, wrap the bottom blanket flap around the floor of the shell and fix it with the timber band to the outside of the studio shell. Then lay out and glue down the whole bottom floor blanket to the bottom panel and screw the bottom timber band.

To do this requires these bands to be positioned individually around its appropriate place and then screwing each of them to the shell of the studio and to each other by their ends at 90 degrees.

At this point there will be the loose blanket flaps protruding from the opening of the studio, ready to be glued down to the visible surface of the polystyrene/acoustic foam coated sheets and UNDER the acoustic foam panels themselves.

This gives your studio opening a tidy, well–presented appearance and makes it an easy space to work within as no polystyrene edges are exposed. Since Polystyrene is easily damaged and will fall apart if your arms or hands rub across any exposed edges, the blanket which covers the edges will prevent any of this accidental damage.

You can now slip in the polystyrene/acoustic foam–lined panels into your studio shell! Weheew! Not far to go now.

Start by holding the top panel in position so the total top panel surface is covered right to the end that will be sealed shut with the tapered end. At this point you might need to do some slight trimming of this panel so that nothing protrudes from any edges.

Now, slip in each side panel, which will actually hold up the top acoustic treated panel. Again, if there is any trimming required of these side panels then do so. Once all three acoustic treated polystyrene panels are positioned and sitting flush against every wall, you can wrap and glue into position the blanket flaps. Once that is done, you can glue down the rest of the acoustic foam panels which were left unglued. Glue them over the top of the blanket flaps.

Now, fasten the end panel of the studio (with its acoustic treated panel already attached), making sure the top panel and the side panels are positioned around the acoustic treated end piece. In fact, the acoustic treated end panel should be tight up against the top and side panels, particularly the top panel, holding the top panel in position.

ATTACHING LID:

This 'lid', or 'door', is another barrier for preventing exterior noises from entering your studio, and is also a way to make your expensive sound equipment secure from theft or from your dearest children's fiddling, possibly causing damage.

The panel reserved and cut out for this should be lined with a thick blanket, much like the floor panel has been. Secure the blanket on with some timber and screws to give it more strength. Mark out where you will be fixing the hinges (based on the same measurements as per earlier explanation) and screw them on to both the the top timber band first (on the 'underside' of the band/longest edge) and then on to the lid itself.

Fasten the lid to the shell when this is complete and re–screw the top timber band to the shell of the studio.

Next, mark out where the locking staples will be screwed on.

Tip: I found that the long part of the locking staples should go on the bottom end of the studio lid, rather than along the bottom of the studio shell itself. And the hasp loop (where the padlocks slip through) should be fixed onto the bottom timber band. This is because as you 'sit' down at your studio, the long staple can scratch your leg as you adjust yourself for the best voicing position, whereas the hoop is much more forgiving with its rounded shape.

STUDIO COMPLETE:

The final result should resemble the following image. If it does, then well done. But there is still a little way to go and that is to make a backboard (easy) and decide how and where you will be placing this studio and equipment (somewhere quiet and as far inside a building and away from outside walls as you can find).

I have had my studio hanging from the rafters on chains and I have also built myself a portable platform with sturdy castors on its feet which is how I am currently using it.

The advantage of suspending your studio from the rafters on a chain is that the space underneath can still be used for storing your lawn mower, or BBQ. However, it will still need to have a platform underneath or next to it for your computer, and it will also need to be secured to something like a wall otherwise it will be wobbly on its chains while you're working within it. I just used a piece of timber which I screwed onto the shell and then to the wall.

I ended up building a platform for my studio out of an old single bed frame and mounted a shelf onto the platform to hold my laptop. It sits underneath my studio and I can move the whole studio around on its rubber castors.

To hang from the rafters, I fixed some metal hooks and eyelets to the sides of my studio and hung chains on a suspension rack – four chains in total from the racks to the studio. I set this up on a pulley system so I could adjust the height of the studio position to suite my working height.

I recommend, if you wish to use a suspension rack that you hang the studio with chains rather than rope because you need the rigidity of chain. You don't want your studio bobbing around as you get overly excited about the latest narration, otherwise the cause and effect movements will be picked up by the microphone and could also damage your expensive audio equipment.

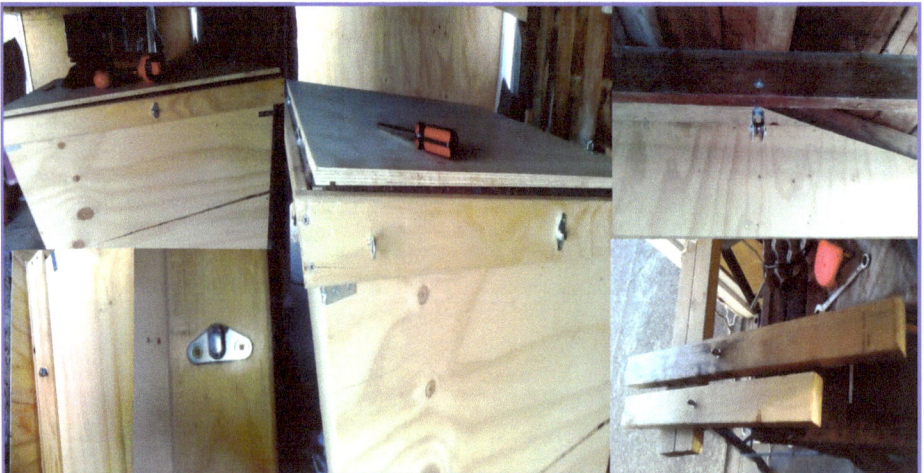

To build your suspension rack, cut to size, according to the studio's front and end widths, 2 timber lengths 100 mm x

50 mm timbers and find their centres. You will need x 3 sturdy eyelets per suspension timber. Position one eyelet per timber in the centre, on top, and two on the bottom at each end, to hook the chains through (with 'S' hooks).

Now, get 2 strong pulley wheels and fix each of them to an appropriate rafter where you want to hang the studio. Make sure that there will be enough space between the studio and the wall to open the studio door so you can get into it.

Next, thread some strong, rigid rope (the less stretchy, the better) through the pulleys to the suspension rack timbers. Tie the ropes to the centre eyelet on the racks and then hook the chains to the four end eyelets. Now you can haul up the ropes, thus lifting the racks and the studio, then tie the ropes off somewhere strong (maybe the rafters?) to hold all the weight.

Tip: **Once you have got your studio all sorted and set up, it's a good idea to throw inside your studio some *Desiccant Silica Gel* sachets to prevent moisture from causing possible mildew and corrosion to your expensive equipment. Every time we speak we are producing moisture and to have it boxed up inside a space, like your studio, could possible create a breeding ground for continual moisture.**

Available from www.dotandtittlepublishing.com

BACK BOARD – Part Three:

This will hold up the lid. It is also an extra barrier, preventing further external noises coming from behind where you will be sitting or standing. It pays to make at least one side board too, to cocoon yourself right inside your studio preventing as much external noise as possible from entering your sound–treated space. I actually ended up screwing on a small, sound treated door to the other side of my studio so I have my studio and workspace totally enclosed.

Tip: One other great thing about this particular Voice–Over Studio design is the fact that, because you are only within your studio from the waist up, it means that you are not likely to get super hot and sweaty like many other studio designs where you are totally inside the entire studio, along with all of your equipment, including the computer.

This part is not difficult. Take the last piece of plywood (1200 mm x 2400 mm) and line one side with thick carpet, blankets or acoustic foams; thus, sound treating it to

prevent waveforms bouncing back towards your microphone within the studio.

Make up some simple feet and legs for the base and stand the plywood up vertically to just above the full height of your studio, making sure you have enough headroom. You can lift the height of the back board by making the legs out of 100 mm x 50 mm timbers and screwing these to the back board, to the height that you require.

You can also find some very thin plywood panels about the same size and create a double–sided wall, lining between the two panels with more of those polystyrene sheets you might have left. This is what I have done. I found them on some old cargo pellets and removed the plywood sheets which was its decking.

I've sandwiched the polystyrene sheets between the thin plywood sheets, positioning the sheets between two vertical timbers (25 mm x 75 mm x 2400 mm) which are running along each vertical edge and then I screwed them all together, putting simple feet on the bottom with a couple of left over timbers lengths. I have lined the inside wall with carpet. When the studio 'door' is opened I then place it on top of this backboard, holding it all firmly into place.

One of my side boards is a brick wall, lined with travel blankets and the other sideboard is simply more of these thick travel blankets draped over the door (which is above my head), and covering the entrance into my work space. It works a charm.

SET–UP:

So, you have made it this far! Brilliant! Now, how do we position all of our equipment and what do we actually need and don't need? This is what this section is about.

The first thing you will need (obviously) is a computer and a microphone. I'll collate a very basic items list for you.

Computer: Something silent. It doesn't matter whether it is one of those fruity ones, or a PC; a laptop or a desktop and it doesn't even have to be the most powerful on the market, with the biggest price attached to it. However, I do strongly recommend a SSD computer, with at least 6 GB of RAM.

SSD's are almost silent and very quick and reliable. In fact, the only noise you would possible hear is the fan and the clicking of the mouse.

Wireless Keyboard and Mouse: Because the computer needs to be positioned outside of the studio – you don't want to have the computer sounds anywhere near your microphone.

Microphone and XLR Cable: This can be a dynamic microphone, or a condenser microphone. I strongly recommend a condenser because they are very sensitive and can gather all of your vocal dynamic range. In other words, they will give you a greater spectrum of audio dynamics to work with and to tweak if you need to, before you deliver it to the ultimate end of the chain... those who are paying you.

Condenser microphone's will also reduce background muddiness as you won't need to turn up the gain on the USB interface as much as you would with a dynamic microphone. This necessary adjustment will only pick up more of the bane of a narrators nightmare; that is... *hiss!* Yes, condenser microphones will pick up more sound/noise, with greater clarity; but this is the whole idea of having a sound proof, sound treated audio studio... So you can produce an original and cleanest sound possible, right at the start! My personal recommendation for a microphone is the RODE NT1 – a well priced, professional microphone.

The XLR cable is to connect the microphone to the USB interface. It's a good idea to have a couple of these because you never know when you might like to introduce a second microphone (like a guest podcast) into your productions.

Pop Filter and Shock Mount: The pop filter stops 'plosives' from 'P's' and 'T's entering violently into your microphone, thus causing terrible 'pops' which will destroy your recordings. The shock mount is necessary with a condenser microphone (but not for a dynamic microphone) because it will prevent any slight movement, bumps and knocks from spoiling your recordings.

USB Interface with Phantom Power: This is what you connect your microphone to your computer with... You will *not* be using a USB microphone, right? Just saying, controversial as it is but if you want to be a professional narrator you will need to start off on a professional playing field; even if you're an entry–level professional. Besides, a USB interface will give you better control of what you will be feeding from the microphone, to the computer. The Phantom Power is necessary to power–up the condenser microphone. My recommendation is the Focusrite Scarlett USB Interface.

Microphone Stand: A must have. Something that can hold your microphone in a consistent position which you can easily move and adjust as required.

Over–ear Headphones: These can be cheap. Some people even use ear–buds, but I wouldn't recommend those. The one key thing that your headphones MUST NOT be, are Noise Cancelling headphones, or Bluetooth headphones. The reason? Well, if they are Noise Cancelling then you will never hear any hiss, or certain dynamics of your recording that need removing; consequently, when you deliver your less-than-perfect recording to your customer, they may never tell you and they also may never hire you again. And Bluetooth headphones tend to have a 'delayed effect', or 'latency' when you listen back to your recordings which makes it very hard to precisely edit your recordings.

LCD Monitor: This will be inside your studio. Measure the width and height of the monitor before you go and purchase one because you want to make sure it isn't too big to fit inside at a comfortable distance and height from where you will be working. The one I have in my studio is a 22 inch and that works well for me.

Platform (for inside the studio): This is to sit the monitor on. I made this from some of the left over timber and a custom-wood sheet I got from a furniture store 400 mm x 500 mm.

I have also screwed my microphone stand to it. And because it is elevated, I can hide several cables and other bits and pieces underneath it, out of the way.

I made this by cutting from the 100 mm x 50 mm timbers, four short legs; each 100 mm long. Then I measured in from the edge of the custom-wood, a 50 mm hang-over to mark where to glue and screw each of these legs, to the corners. I have left this hang-over because it is easier to work around the platform when you are not hitting any of the four legs with a keyboard, your hand or a clip board. The legs are tucked in underneath, also out of the way.

DAW (Digital Audio Workstation): This is not hardware, but is worth mentioning. There are many available, some are free and some are very expensive. Although the quality of the audio production with the various programmes should all be the same, the main differences between them all is 'features'; but what truly matters is the ease of use and how

YOU like the functionality or user–friendliness of them whichever one you chose to go with. I personally use Tracktion T7 – which is free and I thoroughly recommend it. Many people start off with Audacity – which is also free, but Tracktion 7 is far more superior.

Tip: **Most DAW's offer a free trial period so I suggest you go and try as many out as you would like and find the best one for you. If you don't like whichever one's you try then simply uninstall them and unsubscribe to the suppliers. Easy!**

Dog Clicker: One last suggestion of things you will need... is a dog clicker. Yes, a simple tool but with huge time–saving advantages.

The idea of having one of these is that you will hold it in your hand as you are recording and when you make a mistake (and you will) then you 'click' the device near the microphone and it will cause a big 'spike' in your waveform recording. Once you've clicked the device you will go back to the start of the sentence you were recording before the mistake and then carry on recording, this will keep your flow going.

When you go back to edit the waveform then you will be able to see all of those lovely 'spikes' and will zoom into the beginning of the sentence you were recording before the mistake and simply delete it all out, including the 'spike' you made with the dog clicker. Trust me it will save you hours of hunting for mistakes and then editing them out.

FINAL ADVICE:

There are books galore written about voice acting and audio booth/studio construction, but this book is designed solely to show you how to build your own studio at a very affordable level in the industry. I'd love to see how you get along with it. Visit my Facebook page (under Contact the Author), Like and post something about it there. I'd love to see some photos of your studio!

I am giving you some very basic tips of what you will need, to get started with this fantastic new world.

Equipment can cost you hundreds and thousands of dollars and I do not recommend you go out right away and purchase all the top end expensive microphones, DAW programmes or USB interfaces and cables. Start off where you can afford to and see how it works out for you from there. It may work out well, and it may not. If it doesn't, then you haven't spent all your hard-earned retirement funds. No harm done and you will very likely be able to recoup your losses, with the sale of your equipment and even this fantastic studio you have created.

However, if you do well in this industry then don't camp there. Always work towards advancing yourself, your training, your studio and your equipment. You might end up with a fully-fledged, well kited-out studio with several staff working for you. Now wouldn't that be great? Small steps, small advances and small risks, can eventually lead to great things!

If you are new to this world of opportunity, there is just one book I highly recommend, although I do recommend others too, but this particular one is an absolute must have in your list of things you will need to grow your narration/voice–over career... and it is called *"Making Tracks: A Writer's Guide to Audiobooks (and How to Produce Them)", by J. Danial Sawyer.*

This book helped me tremendously in my early days of narration and I still reference back to it when I need to. You can even read my review on it. It's still showing, and I don't know why he doesn't have many more of them.

I would really appreciate any comments, good or bad, in a review from you. Unless someone tells me how I am doing at this 'wordsmith' business, I will never know. So, if you would... please do a review on this, 'Do it yourself' advice book on: amazon, goodreads, ingramspark, lulu, smashwords, or wherever you purchased it from.

And don't forget to check out some of my other books.

You have no idea how valuable it is to me, what you have to say.

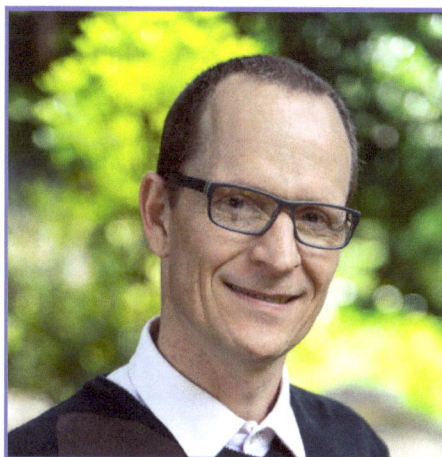

Thank you again

Ian King

Handy Measurements Conversion Chart:

All of these measurements are in the book and so I have specifically converted them from metric to imperial, just because.

Metric measurement first, then **Imperial** conversion of each measurement used in this studio design, follows:

2400 mm = 94.50 Inches

2000 mm = 78.74 Inches

1800 mm = 70.87 Inches

1510 mm = 59.45 Inches

1200 mm = 47.24 Inches

900 mm = 35.43 Inches

884 mm = 34.80 Inches

860 mm = 33.86 Inches

850 mm = 33.47 Inches

836 mm = 32.91 Inches

800 mm = 31.50 Inches

674 mm = 26.54 Inches

650 mm = 25.59 Inches

630 mm = 24.80 Inches

626 mm = 24.65 Inches

602 mm = 23.70 Inches

500 mm = 19.69 Inches

430 mm = 16.93 Inches

400 mm = 15.75 Inches

325 mm = 12.80 Inches

316 mm = 12.44 Inches

300 mm = 11.81 Inches

200 mm = 7.87 Inches

150 mm = 5.91 Inches

100 mm = 3.94 Inches

80 mm = 3.15 Inches

70 mm = 2.76 Inches

50 mm = 1.97 Inches

48 mm = 1.89 Inches

30 mm = 1.18 Inches

25 mm = 0.98 Inches

24 mm = 0.95 Inches

15 mm = 0.60 Inches

12 mm = 0.47 Inches

10 mm = 0.39 Inches

Contact the Author:

Facebook.com/dotandtittle
Goodreads.com/dotandtittle
smashwords.com/profile/view/dotandtittle
Twitter.com/KiwiNarrator

Other Works by This Author:

Books:

The Girl Who Was Buried in Her Ball Gown

(eBook) ISBN: 978–0–473–39751–7
(Paperback) ISBN: 978–0–473–39750–0

Origins Halloween

ISBN: 978–1–311–26866–2

Under the Ocean

(eBook) ISBN: 978–1–3014–2522–8
(Paperback) ISBN: 978–1–4535–6846–0

Meditate Day and Night

ISBN: 978–1–30103–088–0

Graveyard Hill
ISBN: 978-0-473-33004-0

Audiobooks on audible.com:

I may be a an author, but I am also an audio book narrator and you will find my current works in various places, as follows.

Origins Halloween – ISBN: 978-0-473-30337-2

Alexander Turtle III – ASIN: B012O79OEO

Titanic Revisited – ASIN: B0193SJMUI

Any Bloody Idiot – ASIN: B01AO0FD70

Graveyard Hill – ISBN: 978-0-473-33008-8

Toots and Poots in a world full of Snoots – ASIN:B07HNJ7GR2

If you have a novel, or a short story that you'd love to see turned into an audio book and you need a seasoned narrator, then why not consider me! I love audio books and I can do a number of different characters and accents. Check me out at:

https://voice123.com/dotandtittle

Find out more by visiting: www.dotandtittlepublishing.com

or, my narrators fellowship website: www.anznarrators.nz

Please leave a book review :)

© 2019 dotandtittle publishing

https://anznarrators.nz – *Narrators Fellowship*

www.ingramcontent.com/pod-product-compliance
Lightning Source LLC
Chambersburg PA
CBHW041918260326
41914CB00014B/1487